EARTH FILES

RIVERS & LAKES

Chris Oxlade

CONTENTS

Water in the Niagara River plummets 55 metres over the spectacular Niagara Falls between Canada and the USA.

The River Ganges in India is sacred to Hindus. Here, Hindu pilgrims are bathing in the river.

EARTH FILES

RIVERS & LAKES

EARTH FILES – RIVERS AND LAKES
was produced by

David West Children's Books
7 Princeton Court
55 Felsham Road
London SW15 1AZ

Editor: James Pickering
Picture Research: Carrie Haines

First published in Great Britain in 2002 by
Heinemann Library, Halley Court, Jordan Hill, Oxford OX2 8EJ, a division of Reed Educational and Professional Publishing Limited.

OXFORD MELBOURNE AUCKLAND
JOHANNESBURG BLANTYRE GABORONE
IBADAN PORTSMOUTH (NH) USA CHICAGO

06 05 04 03 02
10 9 8 7 6 5 4 3 2 1

ISBN 0 431 15622 0 (HB)
ISBN 0 431 15629 8 (PB)

British Library Cataloguing in Publication Data

Oxlade, Chris
Rivers & lakes. - (Earth Files)
1. Rivers - Juvenile literature
2. Stream ecology - Juvenile literature
I. Title
551.4'83

Printed and bound in Italy

PHOTO CREDITS :
Abbreviations: t-top, m-middle, b-bottom, r-right, l-left, c-centre.

Front cover, 4-5t & 14-15t, 5m, 8-9, 9tr, 13br, 14bl, 15b, 17mr, 20bl, 21ml, bl, bm & br, 24-25t & b, 25tr, 28tr, 29br - Corbis Images. 4-5b & 26-27b (Gavin Hellier), 9tl, 16bl (K. Gillham), 10bl (Gohier), 10-11, 11br (Gina Corrigan), 12tr (S.H. & D.H. Cavanaugh), 12bl (P. Koch), 12br (A. Woolfitt), 16tr (Photri), 19mr (Fred Friberg), 25br (Robert Francis), 26ml (V. Southwell), 27br (Sassoon), 28bl (R. Estall), 7b, 17tr - Robert Harding Picture Library. 11tr, 15tr (Adrian Warren), 16-17b (Su Gooders), 18-19t (Gehan de Silva Wijeyeratne), 20ml (Nick Gordon), 20mr (J.E. Swedberg), 21t (M. Watson), 21mr (P. Morris), 23tr (John Clegg), 27tr (Liz & Tony Bomford), 29mr (C.J. Swedberg), 16mr, 20br - Ardea London Ltd. 18-19m, 21tl - Roger Vlitos. 18-19b, 22-23 (W. Lawler), 20tr (Mike Maidment), 23br (Christine Osborne), 26-27t (M. Whittle), 29ml (Chinch Gryniewicz) - Ecoscene. 22tr (Lauva Sivell) - Papilio.

An explanation of difficult words can be found in the glossary on page 31.

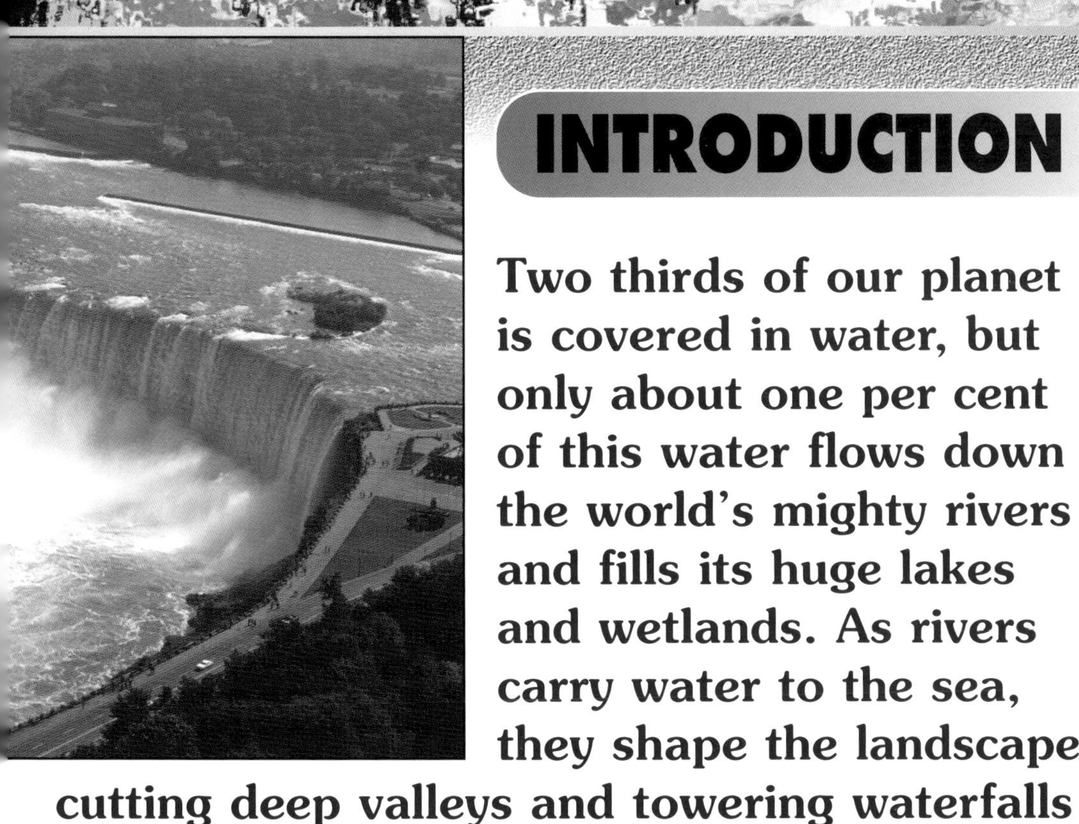

INTRODUCTION

Two thirds of our planet is covered in water, but only about one per cent of this water flows down the world's mighty rivers and fills its huge lakes and wetlands. As rivers carry water to the sea, they shape the landscape, cutting deep valleys and towering waterfalls in one place and creating wide flood plains and vast deltas in others. They bring water to barren deserts and provide habitats for animals and plants.

Elk (called moose in North America) eat plants that grow along lake shores. Their long legs allow them to wade deep into the water to graze.

RIVERS AND LAKES OF THE WORLD

Rivers flow through almost every country in the world. They carry water from the world's mountain ranges, across sweeping plains and down to the sea. They often start or finish in the world's lakes.

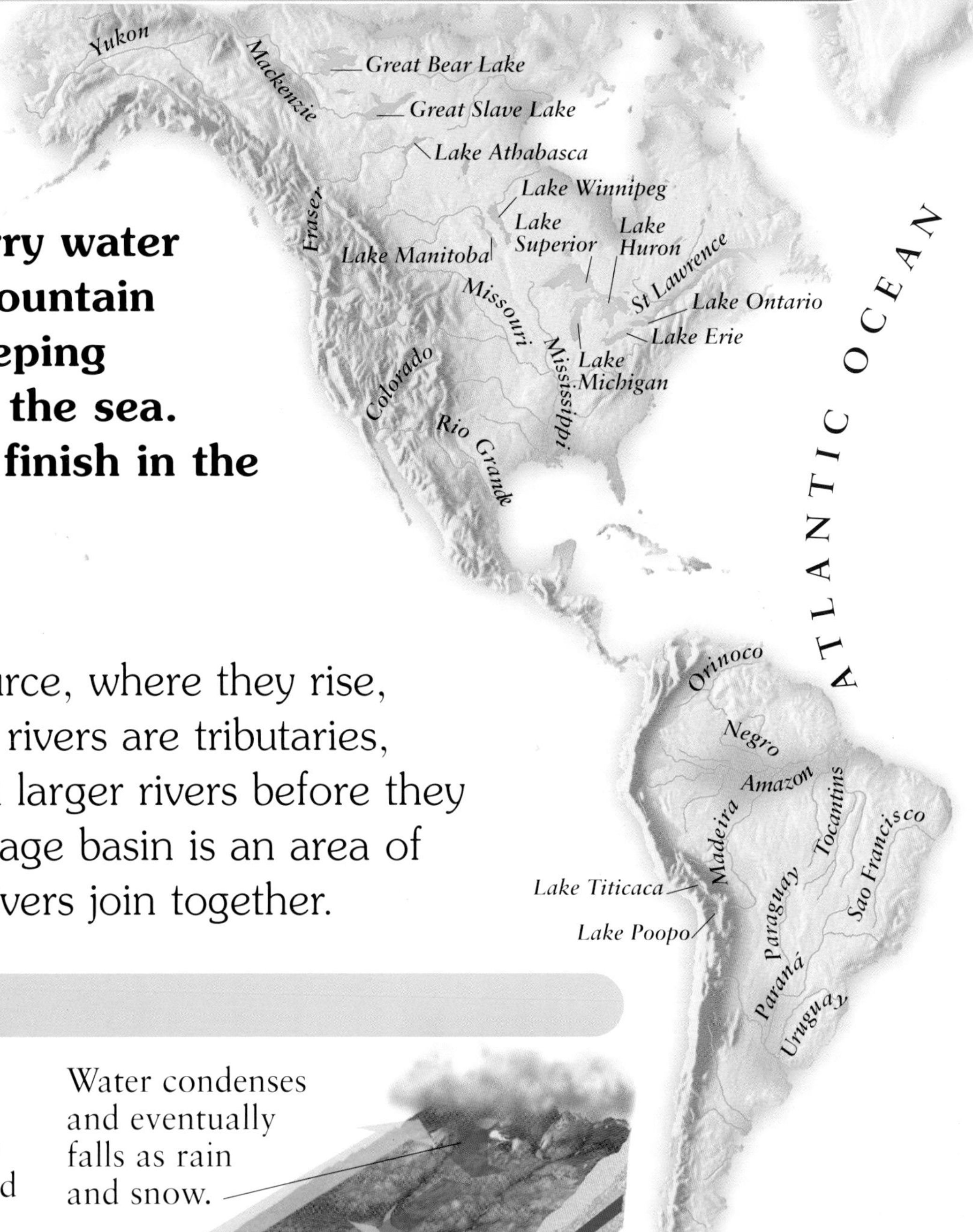

DRAINAGE BASINS

Rivers flow from a source, where they rise, towards the sea. Most rivers are tributaries, which means they join larger rivers before they reach the sea. A drainage basin is an area of land in which all the rivers join together.

THE WATER CYCLE

Water constantly circulates between the oceans, the atmosphere and the land. As it does, it changes from liquid water to water vapour and back. This circulation is called the water cycle. Rivers form part of the cycle. They carry water from the land back to the sea.

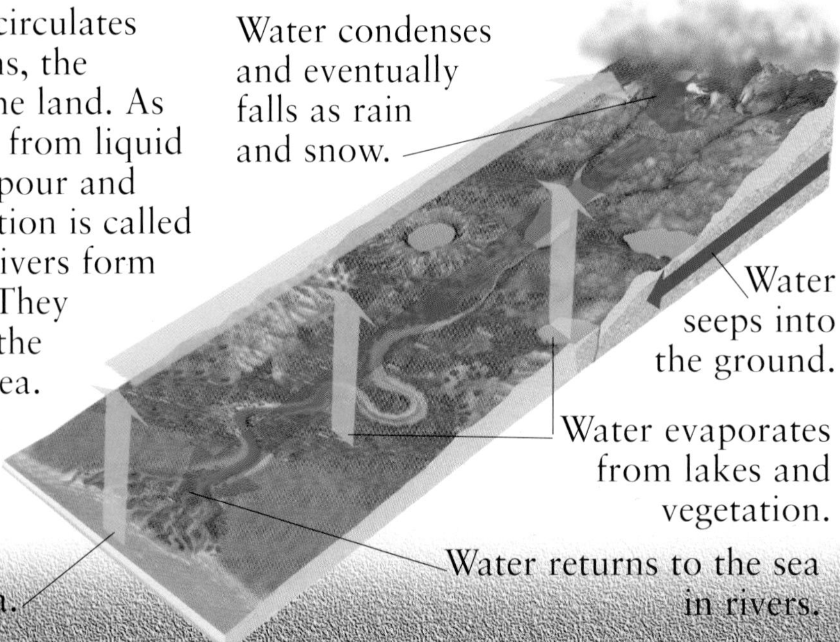

Water condenses and eventually falls as rain and snow.

Water seeps into the ground.

Water evaporates from lakes and vegetation.

Heat from the Sun evaporates water from the sea.

Water returns to the sea in rivers.

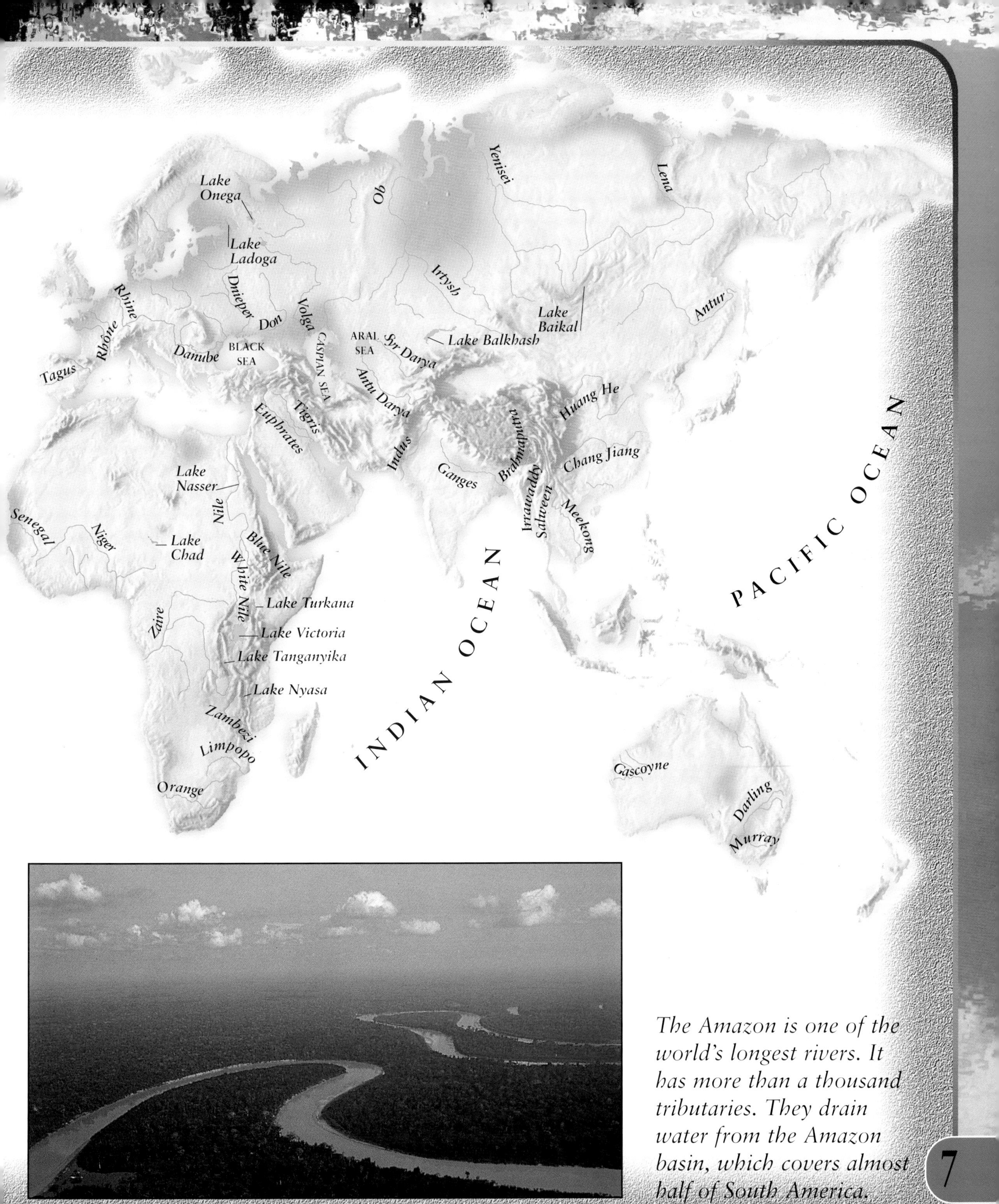

The Amazon is one of the world's longest rivers. It has more than a thousand tributaries. They drain water from the Amazon basin, which covers almost half of South America.

YOUNG RIVERS

A river flows from its source in the hills or mountains, through valleys and plains until it reaches the sea. Geographers divide a river's course into three stages – upper, middle and lower, or young, middle-aged and old.

Part-time rivers

Most rivers flow all year round. Others flow only in winter, or during the rainy season. Others, called ephemeral, or temporary, rivers, flow only after occasional rainstorms. At other times, the river bed is dry or covered in shallow pools.

RIVER SOURCES

The place where a river begins is called its source. The source of most rivers is a patch of high ground where rain-water collects. The water trickles downhill and joins with more water to form a stream. Several streams join to form a young river that flows quickly downhill. Streams and rivers also start where underground water rises to the surface, where glaciers melt, and in lakes.

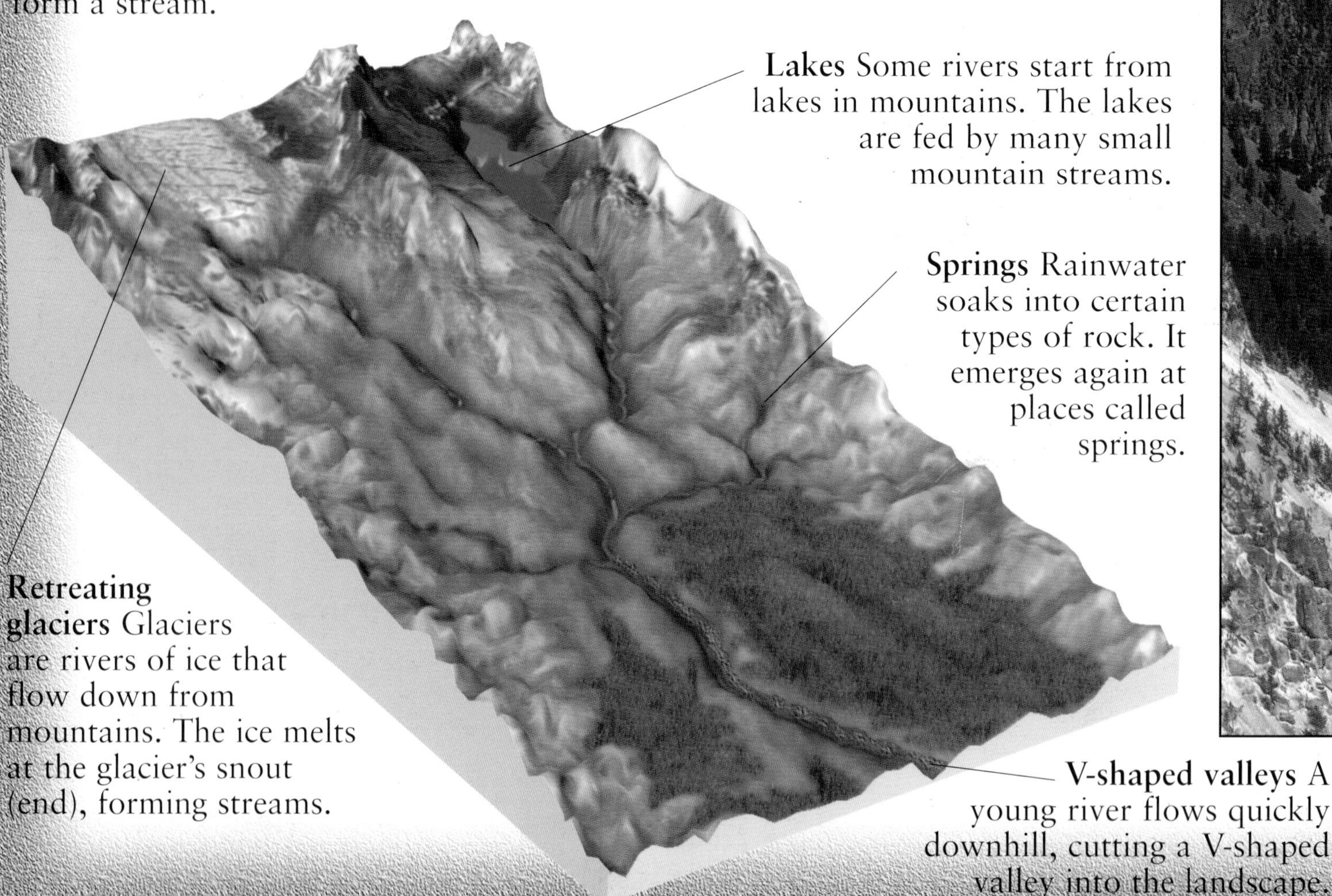

Dry bed of ephemeral river.

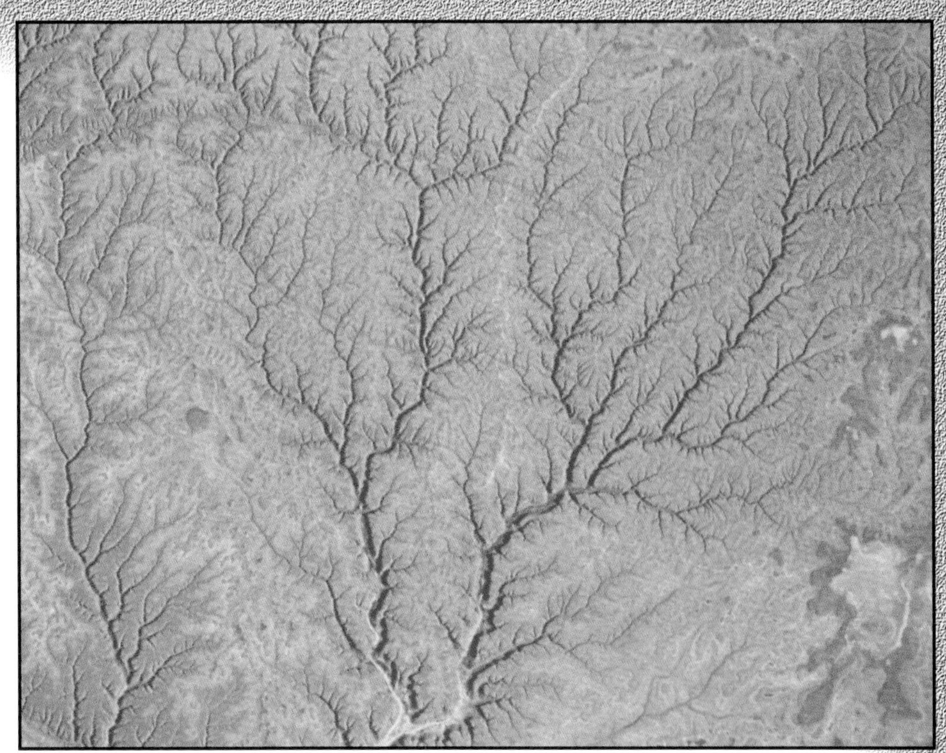

A satellite view of an area of South Yemen. You can see how streams and young rivers join together.

A V-shaped valley cut by the young Yellowstone River in Wyoming, USA.

Fast-flowing rivers

Young rivers flow down from hills or mountains. The land here slopes steeply, and the water is fast flowing, creating rapids and waterfalls. Pieces of rock created by weathering are washed into the river by feeder streams. The pieces range in size from tiny particles to small pebbles and large boulders.

Cutting valleys

The fast-flowing water carries pieces of rock, called the load, along with it. Small particles are carried in the swirling water. Larger ones bump along the river bed, wearing it away. Gradually the river eats into the rock layers, creating a steep-sided, V-shaped valley.

WANDERING RIVERS

After leaving the mountains, a river reaches its middle course. Here the ground is less steep, and the river flows more slowly. It carries more water because it has been joined by more tributaries.

Meanders

Deposition of sediment

Oxbow lake

Middle-aged rivers

Middle-aged rivers flow through wide valleys with flood plains on either side. The river swings, or meanders, from side to side in great bends. In low-lying, flat areas, the river can spread out, creating a swamp or marsh.

The Everglades in Florida, USA, is a huge swamp hundreds of kilometres across.

Sediment from floods creates flood plains.

River divides into many channels in a swamp.

MEANDERS AND OXBOWS

On a river bend, the water flows quickly on the outside, and slowly on the inside. The fast flow erodes (wears away) the outside bank, while sand and silt are deposited by the slow-flowing water on the inside. Gradually the bend widens, forming a meander. During a flood, a river can cut a new channel across the bend, leaving an ox-bow lake.

Oxbow lake, Venezuela.

FLOOD PLAINS

A middle-aged river cannot carry large pieces of rock, but it can carry large amounts of sediment (sand, silt and clay). When a river floods, water spreads across the flat flood plains, and deposits layers of sediment. The regular deposits of fresh sediment make flood plains fertile farming land.

A yellow river

The Huang He is the second largest river in China. It is also known as the Yellow River because of the yellow silt that colours the water. The river picks up the silt as it cuts through layers of soil. The silt is dumped hundreds of kilometres downstream, where the river slows down.

The yellowy waters of the Huang He, China.

TOWARDS THE SEA

As a river nears the sea, it reaches its final stage. The place where it flows into the sea is called its mouth. Here, the fresh river water mixes with the seawater which flows up the river at high tide.

Not all rivers flow into the sea. The Okavango River flows into the Kalahari Desert, forming marshes called the Okavango Delta.

Estuaries

The wide bottom section of a river, where the tide of the sea flows in and out, is called an estuary. The river slows right down, and dumps its load of sand, silt and clay here, forming mud banks.

Rivers do not always form estuaries when they reach the sea. As the Mackenzie River in Canada approaches the coast, many small waterways flow through the flat, frozen landscape to make a delta.

Flood barriers

During very high tides or storm surges, low-lying land next to the tidal section of a river can be flooded by water flowing upstream. The gates of the Thames Barrier are closed to protect low-lying parts of London from the threat of these floods.

Thames flood barriers.

RIVER DELTAS

Where a large river flows quickly into the sea, which often happens during floods, its load of sand, silt and clay is carried out to sea and dumped. Gradually, this sediment builds up to form a fan-shaped area of new land called a delta.

Flood plain
The flat coastal plain is often built up from sediment carried down by the river.

Delta
During floods, the channels change course as land is eroded and built up again.

Farming on delta islands
Deltas are good places to grow crops, but floods make farming hazardous.

Using estuaries and deltas

Many towns and cities have grown up near estuaries and deltas. Estuaries protect the land from storms at sea. Flood plains and deltas are regularly covered with a fresh layer of mineral-rich sediment during floods. This creates rich, fertile land for growing crops.

A satellite view of the Mississippi delta. The sediment is forming new spits of land in the Gulf of Mexico. This type of delta is called a bird's foot delta because of its shape.

WATERFALLS, CAVES AND CANYONS

Flowing across the landscape, rivers create spectacular features such as waterfalls, caves and canyons. These features are formed as the water erodes the rocks that it flows over.

CANYONS

A canyon is a deep valley with very steep sides. A canyon forms when a fast-flowing river cuts down into layers of hard rock. It takes millions of years for a canyon to form. The Grand Canyon in the USA was carved out by the Colorado River.

The Grand Canyon, USA, is up to 1.6 km deep.

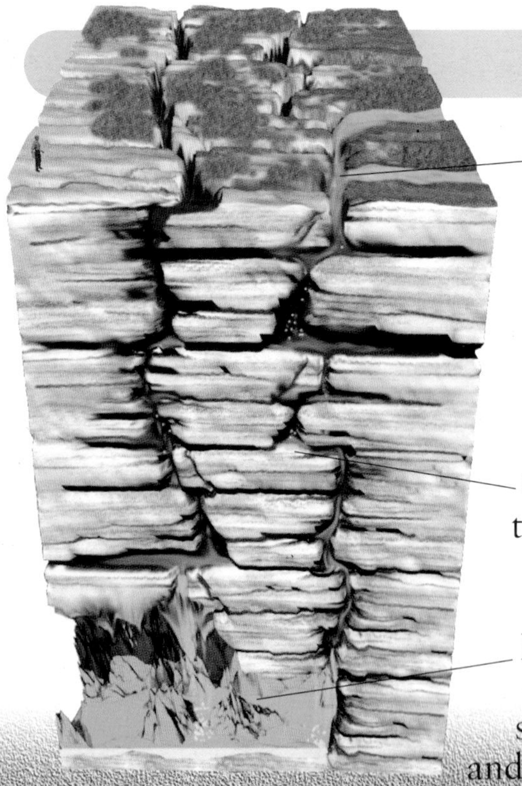

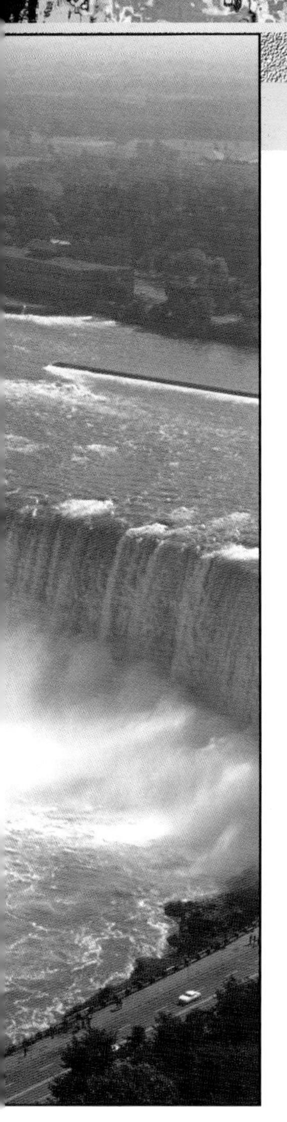

WATERFALL FORMATION

When a river flows over the boundary between hard rock and soft rock, the soft rock erodes quickly, leaving a shelf of hard rock behind. Then the water drops over the shelf, eroding the soft rock even more. This is how a waterfall is formed.

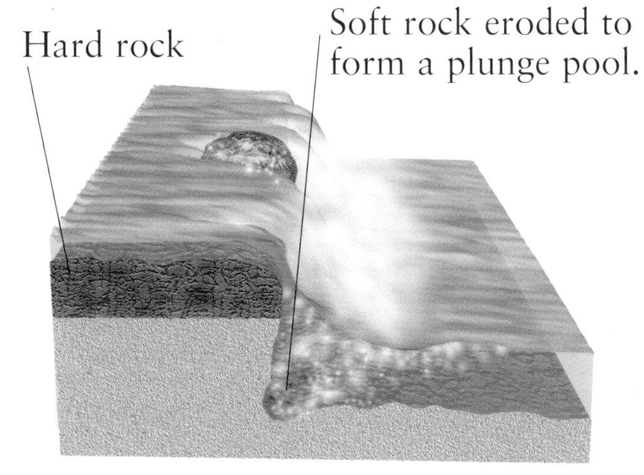

Niagara Falls, on the border between Canada and the USA, is 55 metres high. Erosion is moving the falls more than a metre upstream every year.

The highest falls

Angel Falls, Venezuela, is the highest waterfall on Earth. The water plunges 979 metres as the Rio Churún plunges over the edge of a rocky plateau.

Angel Falls, Venezuela.

CAVE FORMATION

Water cannot flow through most types of rock. It simply flows over it. But limestone is full of tiny holes that allow water to seep inside. Rain-water is slightly acidic. When it flows through the holes, it slowly dissolves the rock, forming systems of caves and tunnels. Rivers flow down the shafts, through the tunnels and emerge further downhill, often many kilometres away.

In a cave, stalactites (hanging from the roof) and stalagmites (on the ground) are formed from the dripping water and limestone solution.

WATERFALLS

Most waterfalls form when a river flows over a rocky shelf (see above). The falls move gradually upstream as the rocky shelf is worn away by the falling water and the edge collapses. Mountain waterfalls are often formed where a section of land on a river's course has dropped away along a fault line.

THE LONGEST RIVER

The world's longest river is the River Nile, in Africa. It flows for 6,695 kilometres, from mountains near Lake Victoria (Nyanza) on the Equator northwards across Africa to its mouth on the Mediterranean Sea.

The Nile Delta is 150 km wide at the sea. Egypt's two largest cities, Cairo and Alexandria, are built on the delta.

Green desert

The Nile has created a green strip of land running through the hot, dry deserts of Egypt and northern Sudan. Over millions of years, annual floods dumped layers of rich silt on the flood plain, creating fertile farmland on each side of the river. More silt built a huge delta, which is also used for growing crops.

Water from the river allows crops to be grown on a narrow strip of land between the River Nile and the barren desert.

Traditional feluccas are used for transport and fishing.

COURSE OF THE NILE

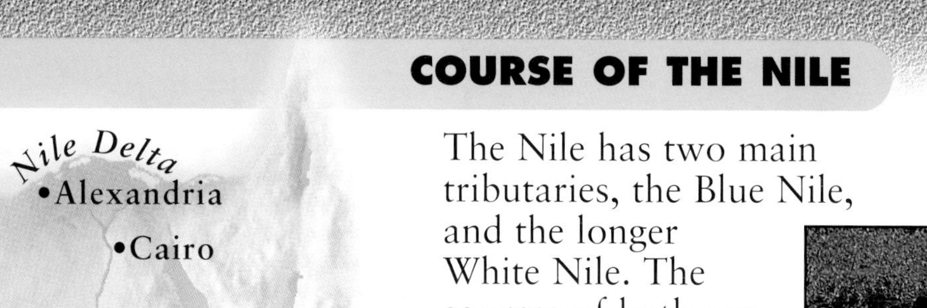

The Nile has two main tributaries, the Blue Nile, and the longer White Nile. The sources of both are in mountainous, rainy areas of Africa. They join at Khartoum in Sudan, flow across the dry Nubian Desert, and into the Mediterranean Sea at the huge Nile Delta.

A farmer raises water into an irrigation channel using an ancient machine called an Archimedean screw.

Lake Nasser.

The Aswan High Dam

The Aswan High Dam, completed in 1971, created Lake Nasser. Its water is used for drinking, irrigating crops and making hydroelectricity. Unfortunately, the dam has caused problems, because it stops rich silt from reaching farmland downstream.

This painting shows ancient Egyptians fishing on the Nile.

WHAT ARE LAKES?

A lake is a hollow in the ground that fills with water from streams and rivers, and rain-water. Lakes range from small, fresh water lakes to vast, salty inland seas. Some lakes are the source of rivers, or have rivers flowing into them.

Lake Nakuru in the Great Rift Valley, Africa, is famous for its pink flamingos.

TYPES OF LAKE

Many lakes form in natural hollows in the landscape. Some lakes form in hollows created by rock movements, such as those in the African Rift Valley.

Here, enormous blocks of rock have sunk down below the surrounding land. Other lakes fill the craters of volcanoes, or hollows gouged out by glaciers. Some lakes form when streams or rivers are blocked by a landslide, icefall or rockfall.

Lake in cirque (hollow) eroded by old glacier.

Crater lake formed by rain falling into crater of volcano.

Lake formed where water flows into natural hollow in the landscape.

Rock layer holding water.

Lake created where water in underground rocks reaches the surface.

Lake in hollow created by rocks slipping down below the surrounding land.

Inland seas

Some large lakes are filled with salt water, rather than fresh water. They are known as inland seas, even though they are not connected to the ocean. The largest inland sea in the world is the Caspian Sea, with an area of 371,000 square kilometres.

Salt lakes

Salt lakes form where water dissolves salts from the lake-bed rocks. Salt lakes sometimes form in deserts, after a rare rainfall. When the lake water evaporates in the sun, it leaves vast salt flats behind.

A tarn in the Welsh mountains. A tarn is a small mountain lake that often forms in a cirque.

Floating in the Dead Sea.

An ephemeral lake, such as this salt lake in New South Wales, Australia, fills when it rains and then evaporates away.

The Dead Sea

The water in the Dead Sea in the Middle East is six times saltier than the oceans. The Dead Sea gets its name because its water is too salty for wildlife to survive. The salt makes the water very dense so that bathers float easily without swimming.

RIVER AND LAKE WILDLIFE

Rivers and lakes, river banks and lake shores provide homes for a huge variety of animals. Other animals come to rivers and lakes to drink, hunt or breed.

The Baikal seal of Lake Baikal, Russia, is the only species of seal that lives in fresh water.

LIFE IN THE WATER

Animals that live in and around rivers and lakes are specially adapted to their habitat. Catfish live in fast-flowing water. They have suckers on their fins for gripping the slippery rocks. Otters have webbed paws and a powerful tail for swimming fast after fish.

River dolphins live in muddy waters where it is difficult to see. They find their way using echolocation and have almost lost their sight.

Beavers build wooden dams across rivers to block off a pool. There they build their dome-shaped homes called lodges from sticks and mud.

Crocodiles eat fish, but also attack larger animals that come to rivers and lakes to drink.

The piranha is just one of the 2,000 species of fish that live in the Amazon River.

Bald eagles and otters both hunt salmon and other fish.

RETURN TO THE RIVER

Most species of salmon hatch from eggs laid in the upper course of a river. As young fish, they swim out to sea for several years, then return to the same river to breed.

Salmon can leap more than 3 metres out of the water. This allows them to swim upstream past rapids and waterfalls.

Axolotl

The strange axolotl only lives in Lake Chalco and Lake Xochimilco in Mexico. It is a type of amphibian called a salamander. It spends its whole life as a tadpole, never turning into an adult.

Axolotl.

Frogs start their lives in ponds as tadpoles.

POND LIFE

The shallow, calm waters of ponds and pools teem with life. Animals such as frogs and mosquitoes come to ponds to lay their eggs.

Dragonfly.

Kingfisher.

RIVER AND LAKE PLANTS

No plants are able to survive in the rushing, tumbling water of a young river. But further downstream, the water is calmer and plants such as rushes and reeds can grow. Similar plants grow around ponds and lakes.

Bulrushes grow in shallow water.

Water plant types

Plants in lakes and slow-flowing rivers are divided into three types. Marginal plants grow in shallow water along river banks or lake shores. Floating plants have leaves and flowers that float on the water surface. Submerged plants grow under the water.

This giant lily grows on the Amazon. Its enormous leaves grow up to 1.5 metres across.

Microscopic plants

The simplest river, lake and pond plants are microscopic algae, such as freshwater desmids. These are eaten by small animals such as insect larvae and tadpoles. It is these plants that make the water look green and slimy.

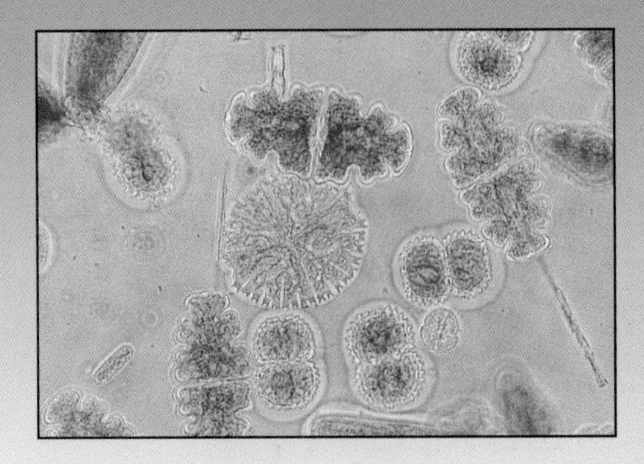

Freshwater desmids.

Mangrove trees in Australia. The roots help to stop the muddy soil from being washed away during floods.

SWAMP PLANTS

Only a few species of plants can survive in the shallow salt water in river estuaries and deltas. One of them is the mangrove, which grows in the tropics. Mangrove trees have arching roots that anchor them in the mud.

On the River Nile, water hyacinths often clog up irrigation channels, but they also absorb pollutants in the water.

HOT SPRINGS AND GEYSERS

Spectacular fountains of steam, hot springs and steaming rivers occur in volcanic areas of the world.

Geysers

Geysers happen in areas of volcanic activity, such as parts of the USA, New Zealand and Iceland. Most famous is the Old Faithful geyser in Yellowstone National Park, USA.

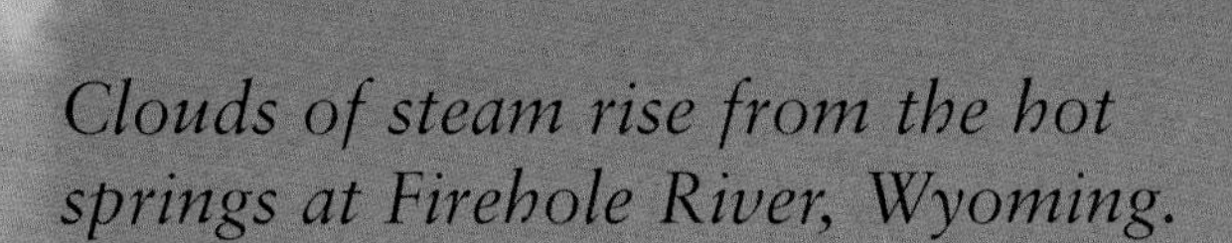

Clouds of steam rise from the hot springs at Firehole River, Wyoming.

HOW GEYSERS WORK

Geysers happen when water flows into an underground chamber that is surrounded by very hot rocks. The water in the chamber is heated until it boils. This creates steam, which fires the water up and out into the air. Then the water dies down, and the process starts again.

Steam

Water seeps into rock.

Hot rocks

Hot springs

A hot spring or thermal spring is a place where water bubbles out of the ground at a high temperature. The water often contains minerals that create strange rock formations as the hot water evaporates.

Old Faithful erupts every hour or so, sending hot water shooting 50 metres high.

Mineral terraces formed by hot springs in the Yellowstone National Park, USA.

Geothermal energy

Most of Iceland's homes and factories are heated and powered by water from hot springs. This type of energy is called geothermal energy. The water is pumped into hot, volcanic rocks and returns as steam, which is fed through turbines to produce electricity. People also bathe in hot geothermal lakes and swimming pools.

Geothermal lake, Iceland.

LIVING BY RIVERS AND LAKES

People who live by rivers and lakes rely on the water for drinking, for washing, for finding food and for transport. Nearly all the world's major cities are built along the shores or mouths of large rivers.

LIFE ON THE WATER

Rivers, lakes and other wetlands dominate the lives of local people. For example, the Turkana nomads of Africa's Rift Valley make their living by catching fish in Lake Turkana. The Marsh Arabs of Iraq rely on the reedbeds for building materials.

Fishing from the bank of the Mekong River in China.

Iraq's Marsh Arabs make their homes from the reeds that grow in the shallow water.

Pilgrims bathing in the Ganges during a festival.

Drying fish on the shores of Lake Malawi. Fishing on the lake is one of Malawi's major industries.

A sacred river

The River Ganges is sacred to Hindus. They believe that bathing in its water washes away their sins. Millions of Hindus make pilgrimages to holy cities along the Ganges' banks, such as Varanasi. When they die, many Hindus have their ashes scattered on the river after being cremated.

Lake Titicaca

High in the Andes between Peru and Bolivia lies Lake Titicaca. The Aymara people who live on the lake shores fish on the lake. They build boats from reeds that grow near the shore.

Reed boat on Lake Titicaca.

USEFUL RIVERS AND LAKES

Rivers and lakes are vital resources. We use their water for drinking and irrigating fields, carrying away waste, transport and leisure activities, such as sailing and fishing.

WATER SUPPLIES

A typical western household uses dozens of litres of water every day for washing, flushing the toilet and so on. The water often comes from rivers, which are dammed to store winter rain.

The massive Glen Canyon Dam, Arizona, USA, forms Lake Powell. Water is drawn from the reservoir for domestic supplies, for industry and for making hydroelectric power.

RIVER TRANSPORT

Goods, especially bulk goods like grain and coal, are easy and cheap to move by water. Waterways, such as the Rhine in Europe and the St Lawrence Seaway in North America, link the oceans to inland ports and cities. Artificial structures such as weirs, locks and canals create level, navigable sections of river and connect lakes and rivers together. A lock allows boats to move between two different levels of a river or canal. **1** The boat enters the lock. **2** The lock gates are closed and sluices in the upstream gates are opened. The water level rises until **3** the gates can be opened and the boat leaves at the higher level.

Locks on the St Lawrence Seaway bypass huge rapids.

Minerals from lakes

Salt lakes contain many dissolved minerals, such as sodium chloride (common salt) and sodium carbonate (soda). The lakes in Africa's Great Rift Valley are rich in soda, and are known as soda lakes. The soda appears when the water evaporates. It is extracted and used in making glass, ceramics, paper and soap.

Soda on the shores of Lake Magadi, Kenya.

River pollution

Raw sewage, rubbish, poisonous chemicals and factory waste are dumped into rivers all over the world. This pollution injures and kills animals and plants that live in the river and further out at sea. It also makes the water too dirty for people living further downstream to use.

Industrial river pollution, USA.

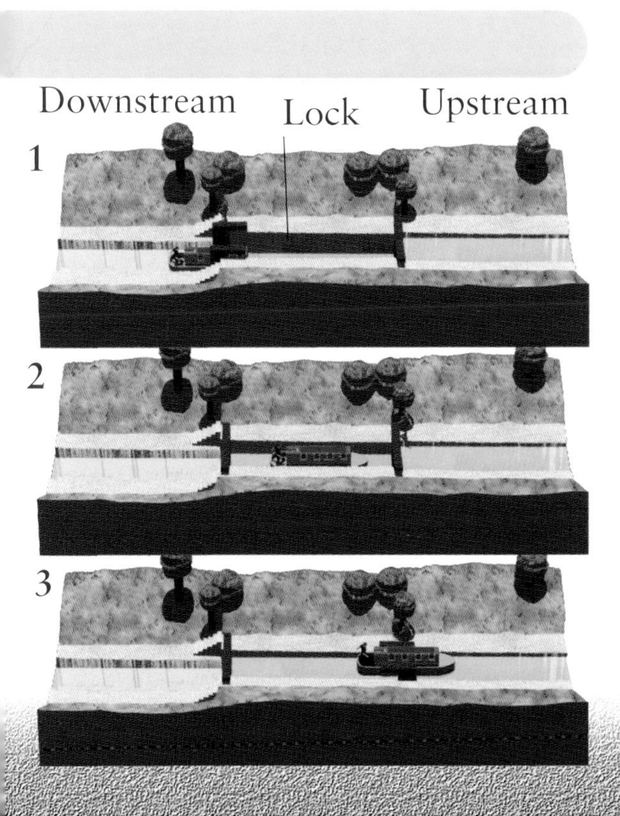

Factories that need huge quantities of water, such as paper mills, are built near rivers.

 AMAZING RIVER AND LAKE FACTS

THE LONGEST RIVER	The River Nile is the longest river in the world. It measures 6,695 kilometres from its source in Burundi, along the White Nile, to its delta on the Mediterranean Sea. Officially, the shortest river is the D River, Oregon, USA, which is just 37 metres long.
THE BIGGEST RIVER	The biggest river in the world, measured by the amount of water that flows down it, is the Amazon. On average 120,000 cubic metres (about 20 swimming pools' worth) of water flows out of its mouth every second. It dumps three million tonnes of sediment into the Atlantic Ocean every day.
THE LARGEST DRAINAGE BASIN	The Amazon basin in South America is the world's largest drainage basin. The Amazon and its tributaries drain an area of 7,050,000 square kilometres between the Andes and the Atlantic Ocean. That is an area three-quarters the size of the continent of Europe.
THE HIGHEST WATERFALL	The Angel Falls in Venezuela is the highest waterfall in the world. Its water falls a total of 979 metres, of which 807 metres is in a single drop. The falls are named after Jimmy Angel, an American pilot who saw them in 1933.
THE LARGEST DELTA	The world's largest river delta is the Ganges-Brahmaputra delta between Bangladesh and West Bengal, India. It covers an area of 75,000 square kilometres, and is more than 400 kilometres wide where it reaches the sea.
THE LARGEST LAKE	The Caspian Sea in Asia is the largest lake in the world. From north to south it measures 1,225 kilometres, and has a surface area of 371,000 square kilometres, the same as Great Britain and Ireland combined. It contains about 90,000 cubic kilometres of salt water.
THE DEEPEST LAKE	The world's deepest lake is Lake Baikal in Russia. At its deepest point the lake bed is 1,637 metres below the surface. It is estimated that Lake Baikal contains about one fifth of all the fresh water in the world.

GLOSSARY

cirque
A bowl-shaped dip in the side of a mountain, created at the start of a glacier.

dense
Having a relatively high density. Density is a measure of how much mass is packed into a certain area.

echolocation
Finding the position of an object by sending out a sound and listening for its echoes bouncing back from the object.

ephemeral
Lasts only for a short period of time.

erosion
The eroding, or wearing away of rocks by the action of wind, rain, flowing water, glaciers and sometimes people.

geothermal
Produced by the heat in the rocks of the Earth's crust.

irrigation
System of supplying water to farmland, made up of channels for the water to flow along.

larvae
Insects that are in the larval stage of their life cycle, between being an egg and turning into an adult.

meander
To swing from side to side in large bends. Also a horseshoe-shaped bend in a meandering river.

sediment
Material such as sand, silt and clay, dropped by a river when it slows down near the sea or flows across flood plains.

tributaries
Small side rivers that flow into a main river.

wetlands
Areas of land that are almost permanently under water, such as marshes and bogs.

INDEX